Homework Helpers
Addition Grade 1

As a parent, you want your child to enjoy learning and to do well in school. The activities in the *Homework Helpers* series will help your child develop the skills and self-confidence that lead to success. Humorous illustrations make these practice activities interesting for your child.

HOW TO USE THIS BOOK

- Provide a quiet, comfortable place to work with your child.

- Plan a special time to work with your child. Create a warm, accepting atmosphere so your child will enjoy spending this time with you. Limit each session to one or two activities.

- Check the answers with your child as soon as an activity has been completed. (Be sure to remove the answer pages from the center of the book before your child uses the book.)

- Topics covered in this book are addition facts, addition of one- and two-digit numbers, addition of money, and addition of three numbers.

- Each section includes one or two pages for reviewing the skills. Time these pages if your child has mastered the skill and is ready to work on improving speed.

Written and Illustrated by Sue Ryono

ISBN #0-86734-121-1

FS-8154 Homework Helpers—Addition Grade 1
All rights reserved—Printed in the U.S.A.
Copyright © 1992 Frank Schaffer Publications, Inc.
23740 Hawthorne Blvd., Torrance, CA 90505

Homework Helper Record

Color the peanut for each page
you complete.

Try These

A.

$$1 + 1 = 2 \qquad 2 + 2 \qquad 2 + 1$$

B.

$$3 + 3 \qquad 0 + 0 \qquad 3 + 1$$

C.

$$4 + 2 \qquad 3 + 2 \qquad 5 + 1 \qquad 1 + 2 \qquad 2 + 3 \qquad 4 + 1$$

D.

$$2 + 2 \qquad 0 + 2 \qquad 0 + 0 \qquad 0 + 6 \qquad 2 + 4 \qquad 1 + 5$$

E.

$$1 + 3 \qquad 6 + 0 \qquad 1 + 4 \qquad 0 + 3 \qquad 1 + 0 \qquad 3 + 2$$

F.

$$2 + 4 \qquad 2 + 3 \qquad 0 + 4 \qquad 4 + 2 \qquad 5 + 0 \qquad 3 + 3$$

FS-8154 Homework Helpers—Addition Grade 1

Fill in the Blanks

I can add!

A. 2 + 1 = _3_   2 + 2 = ___

B. 3 + 3 = ___   1 + 1 = ___

C. 0 + 1 = ___   1 + 2 = ___

D. 1 + 4 = ___   4 + 2 = ___   3 + 2 = ___

E. 2 + 3 = ___   1 + 5 = ___   1 + 3 = ___

F. 2 + 4 = ___   0 + 4 = ___   2 + 2 = ___

G. 4 + 1 = ___   3 + 1 = ___   5 + 1 = ___

H. 0 + 0 = ___   0 + 6 = ___   4 + 2 = ___

I. 3 + 2 = ___   3 + 3 = ___   4 + 0 = ___

J. 5 + 0 = ___   2 + 4 = ___   2 + 3 = ___

Review

A.

5	3	4	2	3	3
+ 1	+ 2	+ 2	+ 2	+ 3	+ 1

B.

0	0	1	2	0	2
+ 0	+ 5	+ 1	+ 3	+ 1	+ 0

C.

0	3	3	2	1	4
+ 6	+ 3	+ 2	+ 4	+ 4	+ 2

D.

0	1	4	1	0	1
+ 6	+ 3	+ 1	+ 0	+ 2	+ 5

E.

4	2	1	2
+ 2	+ 3	+ 5	+ 4

Score: _____

Time: _____

3

You Can Do It

I like math.

A.

$$4 + 4 = 8$$ $$5 + 2$$ $$3 + 5$$

B.

$$4 + 2$$ $$6 + 2$$ $$4 + 3$$

C.

$$3 + 4$$ $$2 + 6$$ $$3 + 2$$ $$5 + 3$$ $$0 + 7$$ $$2 + 5$$

D.

$$6 + 1$$ $$3 + 5$$ $$2 + 4$$ $$3 + 3$$ $$2 + 2$$ $$1 + 6$$

E.

$$4 + 4$$ $$7 + 1$$ $$2 + 5$$ $$5 + 3$$ $$3 + 4$$ $$2 + 6$$

F.

$$5 + 2$$ $$1 + 7$$ $$6 + 2$$ $$8 + 0$$ $$0 + 7$$ $$4 + 3$$

Can You Do This?

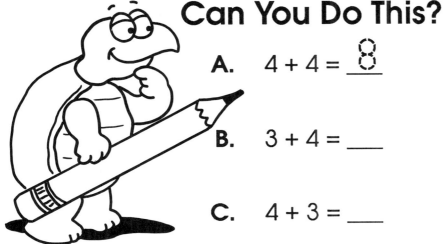

A. $4 + 4 = \underline{8}$ $3 + 5 = \underline{}$

B. $3 + 4 = \underline{}$ $1 + 6 = \underline{}$

C. $4 + 3 = \underline{}$ $5 + 3 = \underline{}$

D. $1 + 7 = \underline{}$ $0 + 8 = \underline{}$ $4 + 2 = \underline{}$

E. $3 + 2 = \underline{}$ $2 + 5 = \underline{}$ $2 + 6 = \underline{}$

F. $3 + 4 = \underline{}$ $3 + 5 = \underline{}$ $5 + 2 = \underline{}$

G. $5 + 3 = \underline{}$ $6 + 2 = \underline{}$ $4 + 4 = \underline{}$

H. $6 + 1 = \underline{}$ $2 + 4 = \underline{}$ $7 + 1 = \underline{}$

I. $2 + 5 = \underline{}$ $0 + 8 = \underline{}$ $6 + 2 = \underline{}$

J. $2 + 6 = \underline{}$ $4 + 3 = \underline{}$ $5 + 2 = \underline{}$

Adding Three Numbers

This is tricky.

A.

2	3	2	1
3	1	0	3
+ 2	+ 3	+ 1	+ 4
7			

B.

2	5	2	2
4	0	2	1
+ 1	+ 3	+ 2	+ 3

C.

3	0	1	1	3	2
0	5	6	3	3	4
+ 5	+ 2	+ 1	+ 2	+ 2	+ 0

D.

3	4	2	2	2	3
2	0	4	3	6	1
+ 3	+ 4	+ 1	+ 3	+ 0	+ 1

E.

3	2	2	2	3	3
2	4	0	0	1	1
+ 2	+ 2	+ 5	+ 2	+ 4	+ 2

Review

A. 4 + 4 = ___   3 + 4 = ___   6 + 2 = ___

B. 0 + 7 = ___   5 + 3 = ___   3 + 2 = ___

C. 5 + 2 = ___   3 + 3 = ___   4 + 3 = ___

D. 2 + 2 = ___   4 + 2 = ___   3 + 5 = ___

E. 2 + 6 = ___   6 + 1 = ___   1 + 5 = ___

F. 8 + 0 = ___   1 + 3 = ___   2 + 4 = ___

G. 2 + 3 = ___   1 + 7 = ___   5 + 3 = ___

H. 2 + 5 = ___   3 + 4 = ___

I. 3 + 5 = ___   1 + 6 = ___

Score: _____

Time: _____

7   FS-8154 Homework Helpers—Addition Grade 1

Review

Go for it!

A.

5	0	3	6	4	1
+ 3	+ 7	+ 4	+ 2	+ 4	+ 7

B.

5	4	3	3	4	2
+ 2	+ 2	+ 5	+ 2	+ 3	+ 4

C.

2	3	8	5	3	2
+ 6	+ 4	+ 0	+ 2	+ 3	+ 5

D.

7	1	2	4
+ 1	+ 6	+ 3	+ 3

Score: _____

Time: _____

FS-8154 Homework Helpers—Addition Grade 1

Sums to Ten

A.

5	6	5	1
+ 5	+ 3	+ 4	+ 9
10			

B.

6	2	8	7
+ 4	+ 7	+ 2	+ 3

C.

4	2	1	5	3	7
+ 5	+ 8	+ 8	+ 3	+ 6	+ 2

D.

3	5	6	4	5	3
+ 7	+ 4	+ 2	+ 3	+ 5	+ 5

E.

3	4	5	2	2	4
+ 4	+ 6	+ 2	+ 4	+ 6	+ 5

F.

5	0	4	7	6	8
+ 4	+ 5	+ 2	+ 2	+ 4	+ 0

FS-8154 Homework Helpers—Addition Grade 1

More Sums

A. 5 + 5 = 10   4 + 4 = ___

B. 5 + 4 = ___   6 + 4 = ___

C. 2 + 6 = ___   5 + 3 = ___

D. 7 + 2 = ___   4 + 5 = ___   6 + 3 = ___

E. 4 + 2 = ___   4 + 3 = ___   4 + 6 = ___

F. 3 + 5 = ___   8 + 2 = ___   2 + 7 = ___

G. 3 + 7 = ___   5 + 2 = ___   2 + 8 = ___

H. 3 + 4 = ___   3 + 6 = ___   7 + 3 = ___

I. 1 + 8 = ___   3 + 7 = ___   5 + 5 = ___

J. 2 + 5 = ___   2 + 7 = ___   2 + 4 = ___

               FS-8154 Homework Helpers—Addition Grade 1

Adding Three Numbers

Can you do this?

A.

2	4	3	5
3	1	4	1
+ 4	+ 4	+ 1	+ 4
9			

B.

3	2	4	0
3	1	2	6
+ 2	+ 5	+ 3	+ 4

C.

6	3	5	6	2	6
2	5	0	0	0	0
+ 2	+ 1	+ 3	+ 3	+ 8	+ 2

D.

4	2	4	3	0	5
0	4	0	6	7	1
+ 3	+ 3	+ 0	+ 0	+ 2	+ 1

E.

2	5	1	2	1	5
4	2	8	5	6	1
+ 2	+ 2	+ 1	+ 2	+ 2	+ 3

Review

A. 5 + 4 = ___   5 + 5 = ___   6 + 3 = ___

B. 0 + 9 = ___   6 + 4 = ___   3 + 7 = ___

C. 2 + 7 = ___   9 + 1 = ___   4 + 4 = ___

D. 5 + 2 = ___   4 + 6 = ___   4 + 3 = ___

E. 3 + 4 = ___   1 + 8 = ___   2 + 5 = ___

F. 3 + 6 = ___   7 + 3 = ___   7 + 2 = ___

G. 8 + 2 = ___   5 + 3 = ___   4 + 2 = ___

H. 6 + 2 = ___   2 + 8 = ___

Score: _____

Time: _____

I. 4 + 5 = ___   3 + 5 = ___

Review

A.
$$\begin{array}{r} 8 \\ +1 \\ \hline \end{array}$$
$$\begin{array}{r} 5 \\ +5 \\ \hline \end{array}$$
$$\begin{array}{r} 5 \\ +4 \\ \hline \end{array}$$
$$\begin{array}{r} 6 \\ +4 \\ \hline \end{array}$$
$$\begin{array}{r} 3 \\ +6 \\ \hline \end{array}$$
$$\begin{array}{r} 3 \\ +7 \\ \hline \end{array}$$

B.
$$\begin{array}{r} 1 \\ +9 \\ \hline \end{array}$$
$$\begin{array}{r} 4 \\ +4 \\ \hline \end{array}$$
$$\begin{array}{r} 7 \\ +2 \\ \hline \end{array}$$
$$\begin{array}{r} 4 \\ +3 \\ \hline \end{array}$$
$$\begin{array}{r} 3 \\ +3 \\ \hline \end{array}$$
$$\begin{array}{r} 4 \\ +5 \\ \hline \end{array}$$

C.
$$\begin{array}{r} 6 \\ +3 \\ \hline \end{array}$$
$$\begin{array}{r} 3 \\ +4 \\ \hline \end{array}$$
$$\begin{array}{r} 4 \\ +6 \\ \hline \end{array}$$
$$\begin{array}{r} 0 \\ +9 \\ \hline \end{array}$$
$$\begin{array}{r} 5 \\ +2 \\ \hline \end{array}$$
$$\begin{array}{r} 2 \\ +8 \\ \hline \end{array}$$

D.
$$\begin{array}{r} 7 \\ +3 \\ \hline \end{array}$$
$$\begin{array}{r} 4 \\ +2 \\ \hline \end{array}$$
$$\begin{array}{r} 2 \\ +7 \\ \hline \end{array}$$
$$\begin{array}{r} 3 \\ +7 \\ \hline \end{array}$$
$$\begin{array}{r} 1 \\ +8 \\ \hline \end{array}$$
$$\begin{array}{r} 2 \\ +5 \\ \hline \end{array}$$

E.
$$\begin{array}{r} 6 \\ +4 \\ \hline \end{array}$$
$$\begin{array}{r} 3 \\ +2 \\ \hline \end{array}$$
$$\begin{array}{r} 2 \\ +4 \\ \hline \end{array}$$
$$\begin{array}{r} 8 \\ +2 \\ \hline \end{array}$$

Score: _____

Time: _____

Sums to Twelve

A.

6	5	5
+ 6	+ 6	+ 7
12		

B.

8	9	8
+ 3	+ 2	+ 2

C.

7	3	2	8	3	7
+ 5	+ 9	+ 8	+ 4	+ 8	+ 4

D.

5	6	7	9	5	6
+ 4	+ 6	+ 4	+ 3	+ 3	+ 5

E.

7	6	6	2	4	3
+ 4	+ 2	+ 4	+ 9	+ 5	+ 7

F.

3	4	3	2	9	8
+ 5	+ 7	+ 8	+ 6	+ 1	+ 3

More Practice

A. $5 + 7 = \underline{12}$ $\qquad$ $9 + 3 = \underline{}$

B. $8 + 2 = \underline{}$ $\qquad$ $6 + 6 = \underline{}$

C. $9 + 2 = \underline{}$ $\qquad$ $7 + 3 = \underline{}$

D. $6 + 5 = \underline{}$ $\qquad$ $7 + 4 = \underline{}$ $\qquad$ $8 + 3 = \underline{}$

E. $3 + 8 = \underline{}$ $\qquad$ $7 + 5 = \underline{}$ $\qquad$ $2 + 9 = \underline{}$

F. $3 + 7 = \underline{}$ $\qquad$ $8 + 4 = \underline{}$ $\qquad$ $5 + 7 = \underline{}$

G. $3 + 9 = \underline{}$ $\qquad$ $4 + 6 = \underline{}$ $\qquad$ $7 + 2 = \underline{}$

H. $5 + 4 = \underline{}$ $\qquad$ $5 + 3 = \underline{}$ $\qquad$ $4 + 8 = \underline{}$

I. $6 + 2 = \underline{}$ $\qquad$ $4 + 7 = \underline{}$ $\qquad$ $6 + 4 = \underline{}$

J. $2 + 7 = \underline{}$ $\qquad$ $5 + 6 = \underline{}$ $\qquad$ $4 + 5 = \underline{}$

FS-8154 Homework Helpers—Addition Grade 1

Adding Three Numbers

A.

```
   6       5       6       4
   2       1       1       2
 + 4     + 5     + 3     + 5
 ────
  12
```

B.

```
   3       4       5       3
   5       3       1       6
 + 3     + 5     + 3     + 1
```

C.

```
   4       3       5       3       6       3
   1       2       6       5       3       3
 + 4     + 6     + 1     + 4     + 2     + 3
```

D.

```
   3       2       4       2       5       2
   3       5       2       3       5       3
 + 4     + 3     + 3     + 6     + 2     + 4
```

E.

```
   2       4       6       5       1       2
   4       4       4       2       7       2
 + 4     + 4     + 0     + 4     + 2     + 5
```

A.   3 + 8 = ___   6 + 4 = ___   3 + 9 = ___

B.   8 + 2 = ___   6 + 6 = ___   7 + 4 = ___

C.   8 + 4 = ___   5 + 6 = ___   2 + 8 = ___

D.   3 + 7 = ___   5 + 7 = ___   4 + 8 = ___

E.   4 + 7 = ___   4 + 6 = ___   8 + 3 = ___

F.   7 + 3 = ___   9 + 3 = ___   6 + 5 = ___

G.   5 + 5 = ___   5 + 4 = ___   3 + 5 = ___

H.   7 + 5 = ___   8 + 4 = ___

Score: _____

Time: _____

I.   4 + 5 = ___   7 + 4 = ___

Review

A.

8	3	7	9	5	3
+ 4	+ 8	+ 5	+ 2	+ 6	+ 9

B.

6	7	7	6	4	8
+ 5	+ 3	+ 4	+ 4	+ 8	+ 3

C.

8	5	2	2	4	9
+ 2	+ 7	+ 8	+ 9	+ 7	+ 3

D.

4	5	6	3
+ 6	+ 4	+ 6	+ 7

Score: _____

Time: _____

Sums to 12

FS-8154 Homework Helpers—Addition Grade 1

Adding Money

A.

7¢	5¢	6¢	5¢
+ 5¢	+ 3¢	+ 6¢	+ 6¢
12¢	¢	¢	¢

B.

5¢	8¢	6¢	9¢
+ 5¢	+ 4¢	+ 4¢	+ 3¢
¢	¢	¢	¢

C.

6¢	7¢	4¢	5¢	4¢	7¢
+ 5¢	+ 4¢	+ 6¢	+ 3¢	+ 3¢	+ 3¢
¢	¢	¢	¢	¢	¢

D.

8¢	3¢	3¢	5¢	5¢	6¢
+ 3¢	+ 7¢	+ 9¢	+ 4¢	+ 7¢	+ 3¢
¢	¢	¢	¢	¢	¢

E.

2¢	7¢	4¢	8¢	3¢	4¢
+ 8¢	+ 2¢	+ 3¢	+ 2¢	+ 8¢	+ 7¢
¢	¢	¢	¢	¢	¢

F.

3¢	6¢	5¢	1¢	2¢	6¢
+ 9¢	+ 5¢	+ 3¢	+ 9¢	+ 9¢	+ 5¢
¢	¢	¢	¢	¢	¢

More Money

A. 5¢ + 7¢ = ___ ¢   8¢ + 3¢ = ___ ¢

B. 5¢ + 5¢ = ___ ¢   4¢ + 7¢ = ___ ¢

C. 6¢ + 5¢ = ___ ¢   7¢ + 3¢ = ___ ¢

D. 5¢ + 4¢ = ___ ¢   9¢ + 3¢ = ___ ¢

E. 6¢ + 6¢ = ___ ¢   9¢ + 2¢ = ___ ¢

F. 2¢ + 8¢ = ___ ¢   3¢ + 9¢ = ___ ¢

G. 6¢ + 4¢ = ___ ¢   3¢ + 8¢ = ___ ¢

H. 7¢ + 5¢ = ___ ¢   4¢ + 3¢ = ___ ¢

I. 5¢ + 6¢ = ___ ¢   4¢ + 5¢ = ___ ¢

J. 8¢ + 4¢ = ___ ¢   4¢ + 4¢ = ___ ¢

What Is the Answer?

Think!

A.

$$8 + 5 = 13$$ $$7 + 7$$ $$7 + 6$$

B.

$$9 + 5$$ $$9 + 4$$ $$6 + 9$$

C.

$$6 + 7$$ $$9 + 6$$ $$5 + 7$$ $$6 + 8$$ $$8 + 7$$ $$8 + 6$$

D.

$$7 + 8$$ $$8 + 5$$ $$5 + 6$$ $$4 + 9$$ $$7 + 5$$ $$5 + 9$$

E.

$$6 + 5$$ $$6 + 8$$ $$7 + 6$$ $$6 + 9$$ $$5 + 8$$ $$8 + 7$$

F.

$$9 + 6$$ $$7 + 7$$ $$7 + 8$$ $$6 + 7$$ $$8 + 6$$ $$9 + 5$$

Still Adding!

A. $9 + 6 = \underline{15}$ $8 + 7 = \underline{}$

B. $8 + 5 = \underline{}$ $9 + 5 = \underline{}$

C. $7 + 8 = \underline{}$ $9 + 4 = \underline{}$

D. $8 + 6 = \underline{}$ $7 + 6 = \underline{}$ $8 + 3 = \underline{}$

E. $6 + 7 = \underline{}$ $5 + 7 = \underline{}$ $7 + 7 = \underline{}$

F. $5 + 6 = \underline{}$ $8 + 7 = \underline{}$ $8 + 2 = \underline{}$

G. $8 + 7 = \underline{}$ $8 + 5 = \underline{}$ $8 + 6 = \underline{}$

H. $7 + 5 = \underline{}$ $9 + 6 = \underline{}$ $6 + 9 = \underline{}$

I. $9 + 4 = \underline{}$ $8 + 4 = \underline{}$ $6 + 5 = \underline{}$

J. $9 + 5 = \underline{}$ $7 + 7 = \underline{}$ $9 + 3 = \underline{}$

Three Addends

A.

4	2	3	4
8	7	6	5
+ 1	+ 3	+ 2	+ 4
13			

B.

3	2	2	4
4	2	1	3
+ 5	+ 7	+ 8	+ 6

C.

4	9	4	8	4	5
2	2	5	2	3	5
+ 0	+ 2	+ 3	+ 4	+ 5	+ 5

D.

5	3	4	4	5	5
4	3	4	4	4	3
+ 5	+ 6	+ 4	+ 6	+ 6	+ 2

E.

4	4	5	6	6	6
4	5	3	2	3	3
+ 6	+ 6	+ 6	+ 5	+ 4	+ 5

Review

A. 5 + 7 = ___   9 + 5 = ___   8 + 7 = ___

B. 8 + 6 = ___   7 + 7 = ___   9 + 6 = ___

C. 8 + 5 = ___   7 + 5 = ___   9 + 4 = ___

D. 6 + 9 = ___   7 + 6 = ___   5 + 9 = ___

E. 6 + 8 = ___   7 + 8 = ___   5 + 8 = ___

F. 6 + 9 = ___   4 + 9 = ___   6 + 8 = ___

G. 5 + 7 = ___   6 + 7 = ___   9 + 5 = ___

H. 9 + 6 = ___   7 + 5 = ___

Score: _____

I. 8 + 7 = ___   8 + 6 = ___

Time: _____

24   FS-8154 Homework Helpers—Addition Grade 1

Pull-Out Answers

Page 1
A. 2, 4, 3
B. 6, 0, 4
C. 6, 5, 6, 3, 5, 5
D. 4, 2, 0, 6, 6, 6
E. 4, 6, 5, 3, 1, 5
F. 6, 5, 4, 6, 5, 6

Page 2
A. 3, 4
B. 6, 2
C. 1, 3
D. 5, 6, 5
E. 5, 6, 4
F. 6, 4, 4
G. 5, 4, 6
H. 0, 6, 6
I. 5, 6, 4
J. 5, 6, 5

Page 3
A. 6, 5, 6, 4, 6, 4
B. 0, 5, 2, 5, 1, 2
C. 6, 6, 5, 6, 5, 6
D. 6, 4, 5, 1, 2, 6
E. 6, 5, 6, 6

Page 4
A. 8, 7, 8
B. 6, 8, 7
C. 7, 8, 5, 8, 7, 7
D. 7, 8, 6, 6, 4, 7
E. 8, 8, 7, 8, 7, 8
F. 7, 8, 8, 8, 7, 7

Page 5
A. 8, 8
B. 7, 7
C. 7, 8
D. 8, 8, 6
E. 5, 7, 8
F. 7, 8, 7
G. 8, 8, 8
H. 7, 6, 8
I. 7, 8, 8
J. 8, 7, 7

Page 6
A. 7, 7, 3, 8
B. 7, 8, 6, 6
C. 8, 7, 8, 6, 8, 6
D. 8, 8, 7, 8, 8, 5
E. 7, 8, 7, 4, 8, 6

Page 7
A. 8, 7, 8
B. 7, 8, 5
C. 7, 6, 7
D. 4, 6, 8
E. 8, 7, 6
F. 8, 4, 6
G. 5, 8, 8
H. 7, 7
I. 8, 7

Page 8
A. 8, 7, 7, 8, 8, 8
B. 7, 6, 8, 5, 7, 6
C. 8, 7, 8, 7, 6, 7
D. 8, 7, 5, 7

Page 9
A. 10, 9, 9, 10
B. 10, 9, 10, 10
C. 9, 10, 9, 8, 9, 9
D. 10, 9, 8, 7, 10, 8
E. 7, 10, 7, 6, 8, 9
F. 9, 5, 6, 9, 10, 8

Page 10
A. 10, 8
B. 9, 10
C. 8, 8
D. 9, 9, 9
E. 6, 7, 10
F. 8, 10, 9
G. 10, 7, 10
H. 7, 9, 10
I. 9, 10, 10
J. 7, 9, 6

Page 11
A. 9, 9, 8, 10
B. 8, 8, 9, 10
C. 10, 9, 8, 9, 10, 8
D. 7, 9, 4, 9, 9, 7
E. 8, 9, 10, 9, 9, 9

Page 12
A. 9, 10, 9
B. 9, 10, 10
C. 9, 10, 8
D. 7, 10, 7
E. 7, 9, 7
F. 9, 10, 9
G. 10, 8, 6
H. 8, 10
I. 9, 8

Page 13
A. 9, 10, 9, 10, 9, 10
B. 10, 8, 9, 7, 6, 9
C. 9, 7, 10, 9, 7, 10
D. 10, 6, 9, 10, 9, 7
E. 10, 5, 6, 10

Page 14
A. 12, 11, 12
B. 11, 11, 10
C. 12, 12, 10, 12, 11, 11
D. 9, 12, 11, 12, 8, 11
E. 11, 8, 10, 11, 9, 10
F. 8, 11, 11, 8, 10, 11

Page 15
A. 12, 12
B. 10, 12
C. 11, 10
D. 11, 11, 11
E. 11, 12, 11
F. 10, 12, 12
G. 12, 10, 9
H. 9, 8, 12
I. 8, 11, 10
J. 9, 11, 9

Pull-Out Answers

Page 16
A. 12, 11, 10, 11
B. 11, 12, 9, 10
C. 9, 11, 12, 12, 11, 9
D. 10, 10, 9, 11, 12, 9
E. 10, 12, 10, 11, 10, 9

Page 17
A. 11, 10, 12
B. 10, 12, 11
C. 12, 11, 10
D. 10, 12, 12
E. 11, 10, 11
F. 10, 12, 11
G. 10, 9, 8
H. 12, 12
I. 9, 11

Page 18
A. 12, 11, 12, 11, 11, 12
B. 11, 10, 11, 10, 12, 11
C. 10, 12, 10, 11, 11, 12
D. 10, 9, 12, 10

Page 19
A. 12¢, 8¢, 12¢, 11¢
B. 10¢, 12¢, 10¢, 12¢
C. 11¢, 11¢, 10¢, 8¢,
   7¢, 10¢
D. 11¢, 10¢, 12¢, 9¢,
   12¢, 9¢
E. 10¢, 9¢, 7¢, 10¢,
   11¢, 11¢
F. 12¢, 11¢, 8¢, 10¢,
   11¢, 11¢

Page 20
A. 12¢, 11¢
B. 10¢, 11¢
C. 11¢, 10¢
D. 9¢, 12¢
E. 12¢, 11¢
F. 10¢, 12¢
G. 10¢, 11¢
H. 12¢, 7¢
I. 11¢, 9¢
J. 12¢, 8¢

Page 21
A. 13, 14, 13
B. 14, 13, 15
C. 13, 15, 12, 14, 15, 14
D. 15, 13, 11, 13, 12, 14
E. 11, 14, 13, 15, 13, 15
F. 15, 14, 15, 13, 14, 14

Page 22
A. 15, 15
B. 13, 14
C. 15, 13
D. 14, 13, 11
E. 13, 12, 14
F. 11, 15, 10
G. 15, 13, 14
H. 12, 15, 15
I. 13, 12, 11
J. 14, 14, 12

Page 23
A. 13, 12, 11, 13
B. 12, 11, 11, 13
C. 6, 13, 12, 14, 12, 15
D. 14, 12, 12, 14, 15, 10
E. 14, 15, 14, 13, 13, 14

Page 24
A. 12, 14, 15
B. 14, 14, 15
C. 13, 12, 13
D. 15, 13, 14
E. 14, 15, 13
F. 15, 13, 14
G. 12, 13, 14
H. 15, 12
I. 15, 14

Page 25
A. 15, 13, 15, 13, 14, 14
B. 14, 13, 12, 15, 13, 13
C. 14, 15, 12, 14, 12, 11
D. 13, 12, 14, 12, 14, 15
E. 12, 15, 14, 11

Page 26
A. 14, 16, 17
B. 15, 15, 18
C. 14, 17, 14, 13, 14, 16
D. 15, 16, 15, 13, 13, 15
E. 16, 14, 14, 13, 17, 14
F. 17, 13, 16, 18, 13, 16

Page 27
A. 17, 18
B. 16, 15
C. 16, 17, 15
D. 14, 14, 13
E. 15, 16, 15
F. 17, 15, 18
G. 16, 13, 14
H. 15, 16, 17
I. 15, 16, 15
J. 18, 17, 16

Page 28
A. 14, 11, 8, 9
B. 10, 12, 16, 14
C. 14, 17, 11, 11, 13, 17
D. 13, 14, 14, 13, 17, 14
E. 18, 13, 15, 15, 14, 11

Page 29
A. 16, 16, 15
B. 17, 15, 14
C. 14, 13, 18
D. 17, 14, 17
E. 13, 15, 16
F. 14, 14, 15
G. 13, 16, 13
H. 17, 13
I. 12, 16

Page 30
A. 17, 18, 16, 15, 14, 15
B. 14, 15, 13, 14, 14, 17
C. 15, 13, 14, 17, 15, 16
D. 14, 16, 13, 15, 13, 16
E. 15, 17, 15, 16

Pull-Out Answers

Page 31
A. 87, 78, 77
B. 49, 67, 69
C. 78, 67, 87, 66, 97, 89
D. 89, 89, 98, 99, 94, 87
E. 86, 85, 97, 96, 89, 88
F. 99, 99, 88, 98, 88, 59

Page 32
A. 59, 68, 82
B. 97, 38, 59
C. 78, 89, 49, 58, 57, 68
D. 78, 99, 89, 89, 49, 67
E. 78, 84, 99, 99, 49, 78
F. 86, 59, 66, 87, 97, 89

Page 33
A. 14, 18, 17
B. 11, 19, 12
C. 13, 15, 16, 14, 10, 18
D. 15, 17, 14
E. 11, 13, 12
F. 16, 19, 18
G. 17, 15, 13
H. 14, 16, 19

Page 34
A. 15, 14
B. 16, 15
C. 12, 11
D. 11. 10
E. 13, 12
F. 15, 14
G. 17, 16
H. 19, 18
I. 18, 17
J. 10, 9
K. 16, 15
L. 17, 16
M. 13, 12

Page 35
A. 15, 14
B. 17, 16
C. 18, 17
D. 16, 15
E. 19, 18
F. 15, 14
G. 12, 11
H. 14, 13
I. 17, 16
J. 13, 12

Page 36
A. 15, 15, 14, 16, 17, 18
B. 13, 11, 14, 17, 13, 16
C. 14, 19, 15, 12, 18, 13
D. 12, 12, 16, 11

Page 37
A. 15, 14, 13
B. 17, 16, 15
C. 14, 13, 12
D. 13, 12, 11
E. 18, 17, 16
F. 12, 11, 10
G. 16, 15, 14
H. 19, 18, 17

Page 38
A. 13, 12, 14, 11
B. 17, 16, 10, 11
C. 18, 17, 15, 13, 11, 16
D. 14, 12, 12, 14, 15, 17
E. 16, 17, 15, 16, 13, 15

Page 39
A. 13, 15, 17, 10
B. 9, 11, 12, 14
C. 16, 10, 12, 14, 11, 13
D. 15, 17, 13, 13, 14, 15
E. 15, 10, 12, 11, 14, 8

Page 40
A. 15, 17
B. 11, 13
C. 12, 14
D. 16, 10, 9
E. 17, 14, 11
F. 13, 15, 12
G. 16, 16, 15
H. 14, 13, 17
I. 12, 11, 14
J. 15, 17, 18

Page 41
A. 12, 15
B. 11, 13
C. 15, 10
D. 14, 16
E. 17, 12, 14
F. 10, 11, 13
G. 15, 14, 8
H. 14, 15, 13
I. 9, 11, 10
J. 17, 15, 17

Page 42
A. 39, 69, 99, 59
B. 89, 99, 47, 95
C. 79, 89, 99, 79, 89, 98
D. 77, 87, 97, 99, 95, 97
E. 77, 89, 98, 37, 49, 79

Page 43
A. 17¢, 16¢, 15¢, 15¢
B. 16¢, 14¢, 16¢, 16¢
C. 18¢, 12¢, 11¢, 12¢, 11¢, 10¢
D. 13¢, 11¢, 13¢, 11¢, 12¢, 12¢
E. 12¢, 11¢, 13¢, 12¢, 13¢, 11¢

Pull-Out Answers

Page 44
A. 12
B. 14
C. 12, 12
D. 10, 9
E. 15, 16
F. 14, 16
G. 15, 13
H. 14, 17
I. 13, 13
J. 13, 18

Page 45
A. 2, 6, 10
B. 4, 8, 12
C. 16, 14, 18, 10, 16, 6
D. 12, 4, 12, 0, 14, 18
E. 10, 2, 16, 4, 6, 10
F. 18, 14, 0, 12, 14, 16

Page 46
A. 9, 10, 13, 12
B. 11, 13, 11, 10
C. 12, 10, 13, 12, 15, 16
D. 11, 16, 17, 12, 16, 15
E. 17, 18, 16, 18

Page 47
A. 86, 56, 96, 98
B. 67, 78, 88, 99
C. 97, 98, 96, 93, 96, 79
D. 49, 85, 95, 89, 89, 99
E. 88, 98, 98, 87

Page 48
A. 6, 7
B. 10, 8
C. 8, 6
D. 6, 7, 10
E. 9, 8, 9
F. 7, 10, 7
G. 9, 10, 9
H. 10, 8, 10
I. 9, 9
J. 8, 10

Page 49
A. 11, 12, 13, 17
B. 13, 16, 16, 18
C. 11, 15, 11, 13, 13, 15
D. 14, 17, 12, 12, 14, 11
E. 15, 15, 14, 16

Review

A.

8	8	9	9	7	9
+ 7	+ 5	+ 6	+ 4	+ 7	+ 5

B.

8	7	5	7	6	5
+ 6	+ 6	+ 7	+ 8	+ 7	+ 8

C.

6	6	6	5	8	9
+ 8	+ 9	+ 6	+ 9	+ 4	+ 2

D.

4	4	7	9	6	8
+ 9	+ 8	+ 7	+ 3	+ 8	+ 7

E.

7	7	8	5
+ 5	+ 8	+ 6	+ 6

Score: _____

Time: _____

25

What's the Sum?

I can do these!

A.

$$8 + 6 = 14 \qquad 9 + 7 \qquad 8 + 9$$

B.

$$8 + 7 \qquad 9 + 6 \qquad 9 + 9$$

C.

$$7 + 7 \qquad 9 + 8 \qquad 9 + 5 \qquad 8 + 5 \qquad 8 + 6 \qquad 7 + 9$$

D.

$$9 + 6 \qquad 8 + 8 \qquad 6 + 9 \qquad 7 + 6 \qquad 9 + 4 \qquad 7 + 8$$

E.

$$9 + 7 \qquad 5 + 9 \qquad 6 + 8 \qquad 6 + 7 \qquad 8 + 9 \qquad 7 + 7$$

F.

$$9 + 8 \qquad 4 + 9 \qquad 7 + 9 \qquad 9 + 9 \qquad 5 + 8 \qquad 8 + 8$$

Zap!

Add Across

A. $8 + 9 = 17$ $9 + 9 = \rule{1.5em}{0.4pt}$

B. $9 + 7 = \rule{1.5em}{0.4pt}$ $8 + 7 = \rule{1.5em}{0.4pt}$

C. $8 + 8 = \rule{1.5em}{0.4pt}$ $9 + 8 = \rule{1.5em}{0.4pt}$ $9 + 6 = \rule{1.5em}{0.4pt}$

D. $5 + 9 = \rule{1.5em}{0.4pt}$ $7 + 7 = \rule{1.5em}{0.4pt}$ $8 + 5 = \rule{1.5em}{0.4pt}$

E. $6 + 9 = \rule{1.5em}{0.4pt}$ $7 + 9 = \rule{1.5em}{0.4pt}$ $7 + 8 = \rule{1.5em}{0.4pt}$

F. $8 + 9 = \rule{1.5em}{0.4pt}$ $7 + 8 = \rule{1.5em}{0.4pt}$ $9 + 9 = \rule{1.5em}{0.4pt}$

G. $9 + 7 = \rule{1.5em}{0.4pt}$ $5 + 8 = \rule{1.5em}{0.4pt}$ $9 + 5 = \rule{1.5em}{0.4pt}$

H. $8 + 7 = \rule{1.5em}{0.4pt}$ $8 + 8 = \rule{1.5em}{0.4pt}$ $9 + 8 = \rule{1.5em}{0.4pt}$

I. $9 + 6 = \rule{1.5em}{0.4pt}$ $7 + 9 = \rule{1.5em}{0.4pt}$ $7 + 8 = \rule{1.5em}{0.4pt}$

J. $9 + 9 = \rule{1.5em}{0.4pt}$ $8 + 9 = \rule{1.5em}{0.4pt}$ $8 + 8 = \rule{1.5em}{0.4pt}$

Add Three!

A.

5	5	2	5
4	3	2	2
+ 5	+ 3	+ 4	+ 2
14			

B.

4	3	4	3
5	3	4	4
+ 1	+ 6	+ 8	+ 7

C.

2	3	5	4	4	9
3	5	3	2	4	0
+ 9	+ 9	+ 3	+ 5	+ 5	+ 8

D.

3	3	2	3	4	3
2	3	4	2	4	2
+ 8	+ 8	+ 8	+ 8	+ 9	+ 9

E.

4	3	4	1	4	3
5	3	4	7	4	2
+ 9	+ 7	+ 7	+ 7	+ 6	+ 6

Review

A. 9 + 7 = ___   8 + 8 = ___   9 + 6 = ___

B. 9 + 8 = ___   8 + 7 = ___   7 + 7 = ___

C. 8 + 6 = ___   8 + 5 = ___   9 + 9 = ___

D. 8 + 9 = ___   9 + 5 = ___   8 + 9 = ___

E. 9 + 4 = ___   6 + 9 = ___   7 + 9 = ___

F. 5 + 9 = ___   6 + 8 = ___   7 + 8 = ___

G. 7 + 6 = ___   9 + 7 = ___   5 + 8 = ___

H. 9 + 8 = ___   4 + 9 = ___

I. 7 + 5 = ___   7 + 9 = ___

Score: _____

Time: _____

Review

A.

$$\begin{array}{r} 9 \\ +8 \\ \hline \end{array} \quad \begin{array}{r} 9 \\ +9 \\ \hline \end{array} \quad \begin{array}{r} 9 \\ +7 \\ \hline \end{array} \quad \begin{array}{r} 8 \\ +7 \\ \hline \end{array} \quad \begin{array}{r} 7 \\ +7 \\ \hline \end{array} \quad \begin{array}{r} 9 \\ +6 \\ \hline \end{array}$$

B.

$$\begin{array}{r} 9 \\ +5 \\ \hline \end{array} \quad \begin{array}{r} 7 \\ +8 \\ \hline \end{array} \quad \begin{array}{r} 8 \\ +5 \\ \hline \end{array} \quad \begin{array}{r} 6 \\ +8 \\ \hline \end{array} \quad \begin{array}{r} 5 \\ +9 \\ \hline \end{array} \quad \begin{array}{r} 8 \\ +9 \\ \hline \end{array}$$

C.

$$\begin{array}{r} 6 \\ +9 \\ \hline \end{array} \quad \begin{array}{r} 9 \\ +4 \\ \hline \end{array} \quad \begin{array}{r} 8 \\ +6 \\ \hline \end{array} \quad \begin{array}{r} 8 \\ +9 \\ \hline \end{array} \quad \begin{array}{r} 8 \\ +7 \\ \hline \end{array} \quad \begin{array}{r} 7 \\ +9 \\ \hline \end{array}$$

D.

$$\begin{array}{r} 6 \\ +8 \\ \hline \end{array} \quad \begin{array}{r} 8 \\ +8 \\ \hline \end{array} \quad \begin{array}{r} 4 \\ +9 \\ \hline \end{array} \quad \begin{array}{r} 9 \\ +6 \\ \hline \end{array} \quad \begin{array}{r} 5 \\ +8 \\ \hline \end{array} \quad \begin{array}{r} 9 \\ +7 \\ \hline \end{array}$$

E.

$$\begin{array}{r} 7 \\ +8 \\ \hline \end{array} \quad \begin{array}{r} 9 \\ +8 \\ \hline \end{array} \quad \begin{array}{r} 6 \\ +9 \\ \hline \end{array} \quad \begin{array}{r} 7 \\ +9 \\ \hline \end{array}$$

Score: _____

Time: _____

Double Digits!

Wow!

A.	15	24	56
	+ 72	+ 54	+ 21
	87		

B.	33	45	50
	+ 16	+ 22	+ 19

C.	68	46	53	44	54	64
	+ 10	+ 21	+ 34	+ 22	+ 43	+ 25

D.	73	62	80	60	64	56
	+ 16	+ 27	+ 18	+ 39	+ 30	+ 31

E.	44	62	83	61	73	61
	+ 42	+ 23	+ 14	+ 35	+ 16	+ 27

F.	51	40	30	81	22	24
	+ 48	+ 59	+ 58	+ 17	+ 66	+ 35

More Big Numbers

A.
```
  56        21        72
+  3      + 47      + 10
----
  59
```

B.
```
  93        35        46
+  4      +  3      + 13
```

C.
```
  20        42        13        33        43        43
+ 58      + 47      + 36      + 25      + 14      + 25
```

D.
```
  42        52        31        49        13        42
+ 36      + 47      + 58      + 40      + 36      + 25
```

E.
```
  64        61        61        52        43        33
+ 14      + 23      + 38      + 47      +  6      + 45
```

F.
```
  52        56        54        16         7        80
+ 34      +  3      + 12      + 71      + 90      +  9
```

Do You See the Trick?

Adding 10 is easy.

A.

$$\begin{array}{r} 10 \\ +\ 4 \\ \hline 14 \end{array} \qquad \begin{array}{r} 10 \\ +\ 8 \\ \hline \end{array} \qquad \begin{array}{r} 10 \\ +\ 7 \\ \hline \end{array}$$

B.

$$\begin{array}{r} 10 \\ +\ 1 \\ \hline \end{array} \qquad \begin{array}{r} 10 \\ +\ 9 \\ \hline \end{array} \qquad \begin{array}{r} 10 \\ +\ 2 \\ \hline \end{array}$$

C.

$$\begin{array}{r} 10 \\ +\ 3 \\ \hline \end{array} \quad \begin{array}{r} 10 \\ +\ 5 \\ \hline \end{array} \quad \begin{array}{r} 10 \\ +\ 6 \\ \hline \end{array} \quad \begin{array}{r} 10 \\ +\ 4 \\ \hline \end{array} \quad \begin{array}{r} 10 \\ +\ 0 \\ \hline \end{array} \quad \begin{array}{r} 10 \\ +\ 8 \\ \hline \end{array}$$

D. $10 + 5 = \underline{}$ $10 + 7 = \underline{}$ $10 + 4 = \underline{}$

E. $10 + 1 = \underline{}$ $10 + 3 = \underline{}$ $10 + 2 = \underline{}$

F. $10 + 6 = \underline{}$ $10 + 9 = \underline{}$ $10 + 8 = \underline{}$

G. $10 + 7 = \underline{}$ $10 + 5 = \underline{}$ $10 + 3 = \underline{}$

H. $10 + 4 = \underline{}$ $10 + 6 = \underline{}$ $10 + 9 = \underline{}$

Add Nine or Ten

I see a pattern.

A.

```
  10        9
+  5      + 5
----      ---
  15       14
```

B.

```
  10        9
+  6      + 6
```

C.

```
  10        9
+  2      + 2
```

D.

```
  10        9
+  1      + 1
```

E.

```
  10        9
+  3      + 3
```

F.

```
  10        9
+  5      + 5
```

G.

```
  10        9
+  7      + 7
```

H.

```
  10        9
+  9      + 9
```

I.

```
  10        9
+  8      + 8
```

J.

```
  10        9
+  0      + 0
```

K.

```
  10        9
+  6      + 6
```

L.

```
  10        9
+  7      + 7
```

M.

```
  10        9
+  3      + 3
```

 FS-8154 Homework Helpers—Addition Grade 1

Nine or Ten Again

A. 10 + 5 = ___   9 + 5 = ___

B. 10 + 7 = ___   9 + 7 = ___

C. 10 + 8 = ___   9 + 8 = ___

D. 10 + 6 = ___   9 + 6 = ___

E. 10 + 9 = ___   9 + 9 = ___

F. 5 + 10 = ___   5 + 9 = ___

G. 2 + 10 = ___   2 + 9 = ___

H. 4 + 10 = ___   4 + 9 = ___

I. 7 + 10 = ___   7 + 9 = ___

J. 3 + 10 = ___   3 + 9 = ___

35   FS-8154 Homework Helpers—Addition Grade 1

Review

Adding 9 or 10

A.

10	9	5	10	9	9
+ 5	+ 6	+ 9	+ 6	+ 8	+ 9

B.

4	2	4	8	10	9
+ 9	+ 9	+ 10	+ 9	+ 3	+ 7

C.

9	10	6	9	8	9
+ 5	+ 9	+ 9	+ 3	+ 10	+ 4

D.

3	2	7	9
+ 9	+ 10	+ 9	+ 2

Score: _____

Time: _____

© Frank Schaffer Publications, Inc.   36   FS-8154 Homework Helpers—Addition Grade 1

Look for a Pattern!

A.

10	9	8
+ 5	+ 5	+ 5
15	14	13

B.

10	9	8
+ 7	+ 7	+ 7

C.

10	9	8
+ 4	+ 4	+ 4

D.

10	9	8
+ 3	+ 3	+ 3

E.

10	9	8
+ 8	+ 8	+ 8

F.

10	9	8
+ 2	+ 2	+ 2

G.

10	9	8
+ 6	+ 6	+ 6

H.

10	9	8
+ 9	+ 9	+ 9

Adding Nine

A.
$$9 + 4$$ $$9 + 3$$ $$9 + 5$$ $$2 + 9$$

click click

B.
$$8 + 9$$ $$9 + 7$$ $$1 + 9$$ $$9 + 2$$

C.
$$9 + 9$$ $$9 + 8$$ $$6 + 9$$ $$4 + 9$$ $$2 + 9$$ $$7 + 9$$

D.
$$5 + 9$$ $$3 + 9$$ $$9 + 3$$ $$9 + 5$$ $$9 + 6$$ $$9 + 8$$

E.
$$9 + 7$$ $$8 + 9$$ $$9 + 6$$ $$7 + 9$$ $$9 + 4$$ $$6 + 9$$

Adding Eight

This is hard for me!

A.

$\begin{array}{r} 8 \\ + 5 \\ \hline 13 \end{array}$
$\begin{array}{r} 8 \\ + 7 \\ \hline \end{array}$
$\begin{array}{r} 8 \\ + 9 \\ \hline \end{array}$
$\begin{array}{r} 8 \\ + 2 \\ \hline \end{array}$

B.

$\begin{array}{r} 8 \\ + 1 \\ \hline \end{array}$
$\begin{array}{r} 8 \\ + 3 \\ \hline \end{array}$
$\begin{array}{r} 8 \\ + 4 \\ \hline \end{array}$
$\begin{array}{r} 8 \\ + 6 \\ \hline \end{array}$

C.

$\begin{array}{r} 8 \\ + 8 \\ \hline \end{array}$
$\begin{array}{r} 2 \\ + 8 \\ \hline \end{array}$
$\begin{array}{r} 4 \\ + 8 \\ \hline \end{array}$
$\begin{array}{r} 6 \\ + 8 \\ \hline \end{array}$
$\begin{array}{r} 3 \\ + 8 \\ \hline \end{array}$
$\begin{array}{r} 5 \\ + 8 \\ \hline \end{array}$

D.

$\begin{array}{r} 8 \\ + 7 \\ \hline \end{array}$
$\begin{array}{r} 9 \\ + 8 \\ \hline \end{array}$
$\begin{array}{r} 8 \\ + 5 \\ \hline \end{array}$
$\begin{array}{r} 5 \\ + 8 \\ \hline \end{array}$
$\begin{array}{r} 8 \\ + 6 \\ \hline \end{array}$
$\begin{array}{r} 7 \\ + 8 \\ \hline \end{array}$

E.

$\begin{array}{r} 8 \\ + 7 \\ \hline \end{array}$
$\begin{array}{r} 2 \\ + 8 \\ \hline \end{array}$
$\begin{array}{r} 8 \\ + 4 \\ \hline \end{array}$
$\begin{array}{r} 3 \\ + 8 \\ \hline \end{array}$
$\begin{array}{r} 8 \\ + 6 \\ \hline \end{array}$
$\begin{array}{r} 0 \\ + 8 \\ \hline \end{array}$

Nine Again

I get it now!

A. 9 + 6 = 15   9 + 8 = ___

B. 9 + 2 = ___   9 + 4 = ___

C. 9 + 3 = ___   9 + 5 = ___

D. 9 + 7 = ___   1 + 9 = ___   9 + 0 = ___

E. 8 + 9 = ___   5 + 9 = ___   2 + 9 = ___

F. 4 + 9 = ___   6 + 9 = ___   3 + 9 = ___

G. 7 + 9 = ___   9 + 7 = ___   9 + 6 = ___

H. 5 + 9 = ___   9 + 4 = ___   8 + 9 = ___

I. 9 + 3 = ___   2 + 9 = ___   9 + 5 = ___

J. 6 + 9 = ___   9 + 8 = ___   9 + 9 = ___

Practice With Eight

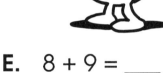

Can you do these?

I think so!

A. 8 + 4 = 12   8 + 7 = ___

B. 8 + 3 = ___   8 + 5 = ___

C. 8 + 7 = ___   8 + 2 = ___

D. 8 + 6 = ___   8 + 8 = ___

E. 8 + 9 = ___   4 + 8 = ___   6 + 8 = ___

F. 2 + 8 = ___   3 + 8 = ___   5 + 8 = ___

G. 7 + 8 = ___   8 + 6 = ___   0 + 8 = ___

H. 6 + 8 = ___   8 + 7 = ___   5 + 8 = ___

I. 8 + 1 = ___   3 + 8 = ___   8 + 2 = ___

J. 9 + 8 = ___   7 + 8 = ___   8 + 9 = ___

Remember This?

Two-digit
numbers!

A. 23   42   61   12
      + 16   + 27   + 38   + 47

B. 33   54   13   72
      + 56   + 45   + 34   + 23

C. 61   52   43   54   45   35
      + 18   + 37   + 56   + 25   + 44   + 63

D. 55   46   37   88   63   44
      + 22   + 41   + 60   + 11   + 32   + 53

E. 53   44   32   20   11   29
      + 24   + 45   + 66   + 17   + 38   + 50

42   FS-8154 Homework Helpers—Addition Grade 1

Money, Money, Money

I like money!

A.   9¢   7¢   8¢   7¢
       + 8¢   + 9¢   + 7¢   + 8¢
       ___¢   ___¢   ___¢   ___¢

B.   9¢   7¢   7¢   8¢
       + 7¢   + 7¢   + 9¢   + 8¢
       ___¢   ___¢   ___¢   ___¢

C.   9¢   5¢   5¢   7¢   6¢   5¢
       + 9¢   + 7¢   + 6¢   + 5¢   + 5¢   + 5¢
       ___¢   ___¢   ___¢   ___¢   ___¢   ___¢

D.   8¢   8¢   5¢   3¢   8¢   4¢
       + 5¢   + 3¢   + 8¢   + 8¢   + 4¢   + 8¢
       ___¢   ___¢   ___¢   ___¢   ___¢   ___¢

E.   9¢   9¢   9¢   3¢   4¢   2¢
       + 3¢   + 2¢   + 4¢   + 9¢   + 9¢   + 9¢
       ___¢   ___¢   ___¢   ___¢   ___¢   ___¢

Add Three Numbers

A. $2 + 6 + 4 = \underline{12}$

B. $3 + 5 + 6 = \underline{}$

C. $4 + 2 + 6 = \underline{}$

$2 + 5 + 5 = \underline{}$

D. $4 + 4 + 2 = \underline{}$

$2 + 6 + 1 = \underline{}$

E. $5 + 4 + 6 = \underline{}$

$4 + 4 + 8 = \underline{}$

F. $4 + 3 + 7 = \underline{}$

$3 + 5 + 8 = \underline{}$

G. $2 + 6 + 7 = \underline{}$

$6 + 1 + 6 = \underline{}$

H. $6 + 3 + 5 = \underline{}$

$3 + 5 + 9 = \underline{}$

I. $1 + 8 + 4 = \underline{}$

$2 + 7 + 4 = \underline{}$

J. $5 + 5 + 3 = \underline{}$

$3 + 6 + 9 = \underline{}$

Doubles!

A.

$$1 + 1$$ $$3 + 3$$ $$5 + 5$$

B.

$$2 + 2$$ $$4 + 4$$ $$6 + 6$$

C.

$$8 + 8$$ $$7 + 7$$ $$9 + 9$$ $$5 + 5$$ $$8 + 8$$ $$3 + 3$$

D.

$$6 + 6$$ $$2 + 2$$ $$6 + 6$$ $$0 + 0$$ $$7 + 7$$ $$9 + 9$$

E.

$$5 + 5$$ $$1 + 1$$ $$8 + 8$$ $$2 + 2$$ $$3 + 3$$ $$5 + 5$$

F.

$$9 + 9$$ $$7 + 7$$ $$0 + 0$$ $$6 + 6$$ $$7 + 7$$ $$8 + 8$$

Review

Three addends!

A.

```
    6        4        4        2
    2        3        4        3
  + 1      + 3      + 5      + 7
```

B.

```
    2        4        4        5
    3        5        5        4
  + 6      + 4      + 2      + 1
```

C.

```
    5        2        5        6        2        2
    2        3        4        2        5        6
  + 5      + 5      + 4      + 4      + 8      + 8
```

D.

```
    5        6        7        5        5        4
    4        3        1        4        3        3
  + 2      +7       + 9      + 3      + 8      + 8
```

E.

```
    4        1        7        4
    5        8        2        5
  + 8      + 9      + 7      + 9
```

Score: _____

Time: _____

Double-Digit Review

A.
```
  70      22      41      23
+ 16    + 34    + 55    + 75
```

I've seen this before.

B.
```
  45      37      28      17
+ 22    + 41    + 60    + 82
```

C.
```
  46      35      24      12      14      14
+ 51    + 63    + 72    + 81    + 82    + 65
```

D.
```
  39      54      72      44      32      20
+ 10    + 31    + 23    + 45    + 57    + 79
```

E.
```
  20      11      32      32
+ 68    + 87    + 66    + 55
```

Score: _____

Time: _____

Review

A. 4 + 2 = ___   4 + 3 = ___

I know these by heart!

B. 2 + 8 = ___   5 + 3 = ___

C. 4 + 4 = ___   2 + 4 = ___

D. 3 + 3 = ___   3 + 4 = ___   5 + 5 = ___

E. 5 + 4 = ___   3 + 5 = ___   4 + 5 = ___

F. 5 + 2 = ___   6 + 4 = ___   2 + 5 = ___

G. 2 + 7 = ___   8 + 2 = ___   6 + 3 = ___

H. 3 + 7 = ___   6 + 2 = ___   4 + 6 = ___

I. 7 + 2 = ___   3 + 6 = ___

Score: _____

J. 2 + 6 = ___   7 + 3 = ___

Time: _____

Review

A.

$$\begin{array}{r} 5 \\ +6 \\ \hline \end{array} \qquad \begin{array}{r} 7 \\ +5 \\ \hline \end{array} \qquad \begin{array}{r} 6 \\ +7 \\ \hline \end{array} \qquad \begin{array}{r} 9 \\ +8 \\ \hline \end{array}$$

Good luck!

B.

$$\begin{array}{r} 7 \\ +6 \\ \hline \end{array} \qquad \begin{array}{r} 7 \\ +9 \\ \hline \end{array} \qquad \begin{array}{r} 8 \\ +8 \\ \hline \end{array} \qquad \begin{array}{r} 9 \\ +9 \\ \hline \end{array}$$

C.

$$\begin{array}{r} 9 \\ +2 \\ \hline \end{array} \qquad \begin{array}{r} 8 \\ +7 \\ \hline \end{array} \qquad \begin{array}{r} 8 \\ +3 \\ \hline \end{array} \qquad \begin{array}{r} 8 \\ +5 \\ \hline \end{array} \qquad \begin{array}{r} 9 \\ +4 \\ \hline \end{array} \qquad \begin{array}{r} 6 \\ +9 \\ \hline \end{array}$$

D.

$$\begin{array}{r} 8 \\ +6 \\ \hline \end{array} \qquad \begin{array}{r} 8 \\ +9 \\ \hline \end{array} \qquad \begin{array}{r} 9 \\ +3 \\ \hline \end{array} \qquad \begin{array}{r} 5 \\ +7 \\ \hline \end{array} \qquad \begin{array}{r} 9 \\ +5 \\ \hline \end{array} \qquad \begin{array}{r} 6 \\ +5 \\ \hline \end{array}$$

E.

$$\begin{array}{r} 7 \\ +8 \\ \hline \end{array} \qquad \begin{array}{r} 9 \\ +6 \\ \hline \end{array} \qquad \begin{array}{r} 5 \\ +9 \\ \hline \end{array} \qquad \begin{array}{r} 9 \\ +7 \\ \hline \end{array}$$

Score: _____

Time: _____

ADDITION AWARD

presented to

_____

for successfully completing
this Homework Helper Book

_____   _____
    signed   **date**

Congratulations!